GENETIC EPISTEMOLOGY

JEAN PIAGET is Professor of Experimental Psychology and Genetic Epistemology at the University of Geneva and for many years has been Director of the International Bureau of Education and of the Institut J. J. Rousseau.

IN THE NORTON LIBRARY

BY JEAN PIAGET

Play, Dreams and Imitation in Childhood
The Origins of Intelligence in Children
The Child's Conception of Number
Genetic Epistemology

BY JEAN PIAGET AND BARBEL INHELDER

The Child's Conception of Space
The Early Growth of Logic in the Child

GENETIC EPISTEMOLOGY

Jean Piaget

TRANSLATED BY ELEANOR DUCKWORTH

The Norton Library
W · W · NORTON & COMPANY · INC ·
NEW YORK

Books That Live

The Norton imprint on a book means that in the publisher's
estimation it is a book not for a single season but for the years.

W. W. Norton & Company, Inc.

ISBN 0 393 00596 8

GENETIC EPISTEMOLOGY

* 1 *

GENETIC EPISTEMOLOGY attempts to explain knowledge, and in particular scientific knowledge, on the basis of its history, its sociogenesis, and especially the psychological origins of the notions and operations upon which it is based. These notions and operations are drawn in large part from common sense, so that their origins can shed light on their significance as knowledge of a somewhat higher level. But genetic epistemology also takes into account, wherever possible, formalization—in particular, logical formalizations applied to equilibrated thought structures and in certain cases to transformations from one level to another in the development of thought.

The description that I have given of the nature of genetic epistemology runs into a major problem, namely, the traditional philosophical view of epistemology. For many philosophers and epistemologists, epistemology is the study of knowledge as it exists at the present moment; it is the analysis of knowledge for its own sake and within its own framework without regard for its development. For these persons, tracing the development of ideas or the development of operations may be of interest to historians or to psychologists but is of no direct concern to epistemologists. This is the major objection to the discipline of genetic epistemology, which I have outlined here.

But it seems to me that we can make the following reply to this objection. Scientific knowledge is in perpetual evolution; it finds itself changed from one day to the next. As a result, we cannot say that on the one hand there is the history of knowledge, and on the other its current state today, as if its current state were somehow definitive or even stable. The current state of knowledge is a moment in history, changing just as rapidly as the state of knowledge in the past has ever changed and, in many instances, more rapidly. Scientific thought, then, is not momentary; it is not a static instance; it is a process. More specifically, it is a process of continual construction and reorganization. This is true in almost every branch of scientific investigation. I should like to cite just one or two examples.

The first example, which is almost taken for granted, concerns the area of contemporary physics or, more specifically, microphysics, where the state of knowledge changes from month to month and certainly alters significantly within the course of a year. These changes often take place even within the work of a single author who transforms his view of his subject matter during the course of his career.

Let us take as a specific instance Louis de Broglie in Paris. A few years ago de Broglie adhered to Niels Bohr's view of indeterminism. He believed with the Copenhagen school that, behind the indeterminism of microphysical events, one could find no determinism, that indeterminism was a very deep reality and that one could even demonstrate the reasons for the necessity of this indeterminism. Well, as it happens, new facts caused de Broglie to change his mind, so that now he maintains the very opposite point of view. So here is one example of transformation in scientific thinking, not over several successive generations but within the career of one creative man of science.

Let us take another example from the area of mathematics. A few years ago the Bourbaki group of mathematicians attempted to isolate the fundamental structures of all mathematics. They established three mother structures: an algebraic structure, a structure of ordering, and a topological structure, on which the structuralist school of mathematics came to be based, and which was seen as the foundation of all mathematical structures, from which all others were derived. This effort of theirs, which was so fruitful, has now been undermined to some extent or at least changed since McLaine and Eilenberg developed the notion of categories, that is, sets of elements taken together, with the set of all functions defined on them. As a result, today part of the Bourbaki group is no longer orthodox but is taking into account the more recent notion of categories. So here is another, rather fundamental area of scientific thinking that changed very rapidly.

Let me repeat once again that we cannot say that on the one hand there is the history of scientific thinking, and on the other the body of scientific thought as it is today; there is simply a continual transformation, a continual reorganization. And this fact seems to me to imply that historical and psychological factors in these changes are of interest in our attempt to understand the nature of scientific knowledge.*

I should like to give one or two examples of areas in which the genesis of contemporary scientific ideas can be understood better in the light of psychological or sociological factors. The first one is Cantor's development of set theory. Cantor developed this theory on the basis of a very fundamental operation, that of one-to-one correspondence. More specifically, by establishing a one-to-one correspondence between the series of whole numbers and the series of even numbers, we obtain a number that is neither a whole number nor an even number but is the first transfinite cardinal number, *aleph zero*. This very elementary operation of one-to-one correspondence, then, enabled Cantor to go beyond the finite number series, which was the only one in use up until his time. Now it is interesting to ask where this opera-

* Another opinion, often quoted in philosophical circles, is that the theory of knowledge studies essentially the question of the validity of science, the criteria of this validity and its justification. If we accept this viewpoint, it is then argued that the study of science as *it is*, as a fact, is fundamentally irrelevant. Genetic epistemology, as we see it, reflects most decidedly this separation of norm and fact, of valuation and description, We believe that, to the contrary, only in the real development of the sciences can we discover the implicit values and norms that guide, inspire, and regulate them. Any other attitude, it seems to us, reduces to the rather arbitrary imposition on knowledge of the personal views of an isolated observer. This we want to avoid.

tion of one-to-one correspondence came from. Cantor did not invent it, in the sense that one invents a radically new construction. He found it in his own thinking; it had already been a part of his mental equipment long before he even turned to mathematics, because the most elementary sort of sociological or psychological observation reveals that one-to-one correspondence is a primitive operation. In all sorts of early societies it is the basis for economic exchange, and in small children we find its roots even before the level of concrete operations. The next question that arises is, what is the nature of this very elementary operation of one-to-one correspondence? And right away we are led to a related question: what is the relationship of one-to-one correspondence to the development of the notion of natural numbers? Does the very widespread presence of the operation of one-to-one correspondence justify the thesis of Russell and Whitehead that number is the class of equivalent classes (equivalent in the sense of one-to-one correspondence among the members of the classes)? Or are the actual numbers based on some other operations in addition to one-to-one correspondence? This is a question that we shall examine in more detail later. It is one very striking instance in which a knowledge of the psychological foundations of a notion has implications for the epistemological understanding of this notion. In studying the development of the notion of number in children we can see whether or not it is based simply on the notion of classes of equivalent classes or whether some other operation is also involved.

I should like to go on now to a second example and to

raise the following question: how is it that Einstein was able to give a new operational definition of simultaneity at a distance? How was he able to criticize the Newtonian notion of universal time without giving rise to a deep crisis within physics? Of course his critique had its roots in experimental findings, such as the Michaelson-Morley experiment—that goes without saying. Nonetheless, if this redefinition of the possibility of events to be simultaneous at great distances from each other went against the grain of our logic, there would have been a considerable crisis within physics. We would have had to accept one of two possibilities: either the physical world is not rational, or else human reason is impotent—incapable of grasping external reality. Well, in fact nothing of this sort happened. There was no such upheaval. A few metaphysicians (I apologize to the philosophers present) such as Bergson or Maritain were appalled by this revolution in physics, but for the most part and among scientists themselves it was not a very drastic crisis. Why in fact was it not a crisis? It was not a crisis because simultaneity is not a primitive notion. It is not a primitive concept, and it is not even a primitive perception. I shall go into this subject further later on, but at the moment I should just like to state that our experimental findings have shown that human beings do not perceive simultaneity with any precision. If we look at two objects moving at different speeds, and they stop at the same time, we do not have an adequate perception that they stopped at the same time. Similarly, when children do not have a very exact idea of what simultaneity is, they do not conceive of it independently of the speed at which objects are traveling. Simultaneity, then, is not a primitive intuition; it is an intellectual construction.

Long before Einstein, Henri Poincaré did a great deal of work in analyzing the notion of simultaneity and revealing its complexities. His studies took him, in fact, almost to the threshold of discovering relativity. Now if we read his essays on this subject, which, by the way, are all the more interesting when considered in the light of Einstein's later work, we see that his reflections were based almost entirely on psychological arguments. Later on I shall show that the notion of time and the notion of simultaneity are based on the notion of speed, which is a more primitive intuition. So there are all sorts of reasons, psychological reasons, that can explain why the crisis brought about by relativity theory was not a fatal one for physics. Rather, it was readjusting, and one can find the psychological routes for this readjustment as well as the experimental and logical basis. In point of fact, Einstein himself recognized the relevance of psychological factors, and when I had the good chance to meet him for the first time in 1928, he suggested to me that is would be of interest to study the origins in children of notions of time and in particular of notions of simultaneity.

What I have said so far may suggest that it can be helpful to make use of psychological data when we are considering the nature of knowledge. I should like now to say that it is more than helpful; it is indispensable. In fact, all epistemologists refer to psychological factors in their analyses, but for the most part their references to psychology are speculative and are not based on psychological research. I am convinced that all epistemology brings up factual problems as well as formal ones, and once factual problems are encountered, psychological findings become relevant and should be taken into account. The unfortunate thing for psychology is that everybody thinks of himself as a psychologist. This is not true for the field of physics, or for the field of philosophy, but it is unfortunately true for psychology. Every man considers himself a psychologist. As a result, when an epistemologist needs to call on some psychological aspect, he does not refer to psychological research and he does not consult psychologists; he depends on his own reflections. He puts together certain ideas and relationships within his own thinking, in his personal attempt to resolve the psychological problem that has arisen. I should like to cite some instances in epistemology where psychological findings can be pertinent, even though they may seem at first sight far removed from the problem.

My first example concerns the school of logical positivism. Logical positivists have never taken psychology into account in their epistemology, but they affirm that logical beings and mathematical beings are nothing but linguistic structures. That is, when we are doing logic or mathematics, we are simply using general syntax, general semantics, or general pragmatics in the sense of Morris, being in this case a rule of the uses of language in general. The position in general is that logical and mathematical reality is derived from language. Logic and mathematics are nothing but specialized linguistic structures. Now here it becomes pertinent to examine factual findings. We can look to see whether there is any logical behavior in children before language develops. We can look to see whether the coordinations of their actions reveal a logic of classes, reveal an ordered system, reveal correspondence structures. If indeed we find logical structures in the coordinations of actions in small children even before the development of language, we are not in a position to say that these logical structures are derived from language. This is a question of fact and should be approached not by speculation but by an experimental methodology with its objective findings.

The first principle of genetic epistemology, then, is this— to take psychology seriously. Taking psychology seriously means that, when a question of psychological fact arises, psychological research should be consulted instead of trying to invent a solution through private speculation.

It is worthwhile pointing out, by the way, that in the field of linguistics itself, since the golden days of logical positivism, the theoretical position has been reversed. Bloomfield in his time adhered completely to the view of the logical positivists, to this linguistic view of logic. But currently, as you know, Chomsky maintains the opposite position. Chomsky asserts, not that logic is based on and derived from language, but, on the contrary, that language is based on logic, on reason, and he even considers this reason to be innate. He is perhaps going too far in maintaining that it is innate; this is once again a question to be decided by referring to facts, to research. It is another problem for the field of psychology to determine. Between the rationalism that Chomsky is defending nowadays (according to which language is based on reason, which is thought to be innate in man) and the linguistic view of the positivists (according to which logic is simply a linguistic convention), there is a whole selection of possible solutions, and the choice among these solutions must be made on the basis of fact, that is, on the basis of psychological research. The problems cannot be resolved by speculation.

I do not want to give the impression that genetic epistemology is based exclusively on psychology. On the contrary, logical formalization is absolutely essential every time that we can carry out some formalization; every time that we come upon some completed structure in the course of the development of thought, we make an effort, with the collaboration of logicians or of specialists within the field that we are considering, to formalize this structure. Our hypothesis is that there will be a correspondence between the psychological formation on the one hand, and the formalization on the other hand. But although we recognize the importance of formalization in epistemology, we also realize that formalization cannot be sufficient by itself. We have been attempting to point out areas in which psychological experimentation is indispensable to shed light on certain epistemological problems, but even on its own grounds there are a number of reasons why formalization can never be sufficient by itself. I should like to discuss three of these reasons.

The first reason is that there are many different logics, and not just a single logic. This means that no single logic is strong enough to support the total construction of human knowledge. But it also means that, when all the different logics are taken together, they are not sufficiently coherent with one another to serve as the foundation for human knowledge. Any one logic, then, is too weak, but all the logics taken together are too rich to enable logic to form a single value basis for knowledge. That is the first reason why formalization alone is not sufficient.

The second reason is found in Gödel's theorem. It is the fact that there are limits to formalization. Any consistent system sufficiently rich to contain elementary arithmetic cannot prove its own consistency. So the following questions arise: logic is a formalization, an axiomatization of something, but of what exactly? What does logic formalize? This is a considerable problem. There are even two problems here. Any axiomatic system contains the undemonstrable propositions or the axioms, at the outset, from which the other propositions can be demonstrated, and also the undefinable, fundamental notions on the basis of which the other notions can be defined. Now in the case of logic what lies underneath the undemonstrable axioms and the undefinable notions? This is the problem of structuralism in logic, and it is a problem that shows the inadequacy of formalization as the fundamental basis. It shows the necessity for considering thought itself as well as considering axiomatized logical systems, since it is from human thought that the logical systems develop and remain still intuitive.

The third reason why formalization is not enough is that epistemology sets out to explain knowledge as it actually is within the areas of science, and this knowledge is, in fact not purely formal: there are other aspects to it. In this context I should like to quote a logician friend of mine, the late Evert W. Beth. For a very long time he was a strong adversary of psychology in general and the introduction of psychological observations into the field of epistemology, and by that token an adversary of my own work, since my work was based on psychology. Nonetheless, in the interests of an intellectual confrontation, Beth did us the honor of coming to one of our symposia on genetic epistemology and looking more closely at the questions that were concerning us. At the end of the symposium he agreed to co-author with me, in spite of his fear of psychologists, a work that we called *Mathematical and Psychological Epistemology*. This has appeared in French and is being translated into English. In his conclusion to this volume, Beth wrote as follows: "The problem of epistemology is to explain how real human thought is capable of producing scientific knowledge. In order to do that we must establish a certain coordination between logic and psychology." This declaration does not suggest that psychology ought to interfere directly in logic—that is of course not true—but it does maintain that in epistemology both logic and psychology should be taken into account, since it is important to deal with both the formal aspects and the empirical aspects of human knowledge.

So, in sum, genetic epistemology deals with both the formation and the meaning of knowledge. We can formulate our problem in the following terms: by what means does the human mind go from a state of less sufficient knowledge to a state of higher knowledge? The decision of what is lower or less adequate knowledge, and what is higher knowledge, has of course formal and normative aspects. It is not up to psychologists to determine whether or not a certain state of knowledge is superior to another state. That decision is one for logicians or for specialists within a given realm of science. For instance, in the area of physics, it is up to physicists to decide whether or not a given theory shows some progress over another theory. Our problem, from the point of view of psychology and from the point of view of genetic epistemology, is to explain how the transition is made from a lower level of knowledge to a level that is judged to be higher. The nature of these transitions is a factual question. The transitions are historical or psychological or sometimes even biological, as I shall attempt to show later.

The fundamental hypothesis of genetic epistemology is that there is a parallelism between the progress made in the logical and rational organization of knowledge and the corresponding formative psychological processes. Well, now, if that is our hypothesis, what will be our field of study? Of course the most fruitful, most obvious field of study would be reconstituting human history—the history of human thinking in prehistoric man. Unfortunately, we are not very well informed about the psychology of Neanderthal man or about the psychology of *Homo siniensis* of Teilhard de Chardin. Since this field of biogenesis is not available to us, we shall do as biologists do and turn to ontogenesis. Nothing could be more accessible to study than the ontogenesis of these notions. There are children all around us. It is with children that we have the best chance of studying the development of logical knowledge, mathematical knowledge, physical knowledge, and so forth. These are the things that I shall discuss later in the book.

So much for the introduction to this field of study. I should like now to turn to some specifics and to start with the development of logical structures in children. I shall begin by making a distinction between two aspects of thinking that are different, although complementary. One is the figurative aspect, and the other I call the operative aspect. The figurative aspect is an imitation of states taken as momentary and static. In the cognitive area the figurative functions are, above all, perception, imitation, and mental imagery, which is in fact interiorized imitation. The operative aspect of thought deals not with states but with transformations from one state to another. For instance, it includes actions themselves, which transform objects or states, and it also includes the intellectual operations, which are essentially systems of transformation. They are actions that are comparable to other actions but are reversible, that is, they can be carried out in both directions (this means that the results of action A can be eliminated by another action B, its inverse: the product of A with B leading to the identity operation, leaving the state unchanged) and are capable of being interiorized; they can be carried out through representation and not through actually being acted out. Now, the figurative aspects are always subordinated to the operative aspects. Any state can be understood only as the result of certain transformations or as the point of departure for other transformations. In other words, to my way of thinking the essential aspect of thought is its operative and not its figurative aspect.

To express the same idea in still another way, I think that human knowledge is essentially active. To know is to assimilate reality into systems of transformations. To know is to transform reality in order to understand how a certain state is brought about. By virtue of this point of view, I find myself opposed to the view of knowledge as a copy, a passive copy, of reality. In point of fact, this notion is based on a vicious circle: in order to make a copy we have to know the model that we are copying, but according to this theory of knowledge the only way to know the model is by copying it, until we are caught in a circle, unable ever to know whether our copy of the model is like the model or not. To my way of thinking, knowing an object does not mean copying it— it means acting upon it. It means constructing systems of transformations that can be carried out on or with this object. Knowing reality means constructing systems of transformations that correspond, more or less adequately, to reality. They are more or less isomorphic to transformations of reality. The transformational structures of which knowledge consists are not copies of the transformations in reality; they are simply possible isomorphic models among which experience can enable us to choose. Knowledge, then, is a system of transformations that become progressively adequate.

It is agreed that logical and mathematical structures are abstract, whereas physical knowledge—the knowledge based on experience in general—is concrete. But let us ask what logical and mathematical knowledge is abstracted from. There are two possibilities. The first is that, when we act upon an object, our knowledge is derived from the object itself. This is the point of view of empiricism in general, and it is valid in the case of experimental or empirical knowledge for the most part. But there is a second possibility: when we are acting upon an object, we can also take into account the action itself, or operation if you will, since the transformation can be carried out mentally. In this hypothesis the abstraction is drawn not from the object that is acted upon, but from the action itself. It seems to me that this is the basis of logical and mathematical abstraction.

In cases involving the physical world the abstraction is abstraction from the objects themselves. A child, for instance, can heft objects in his hands and realize that they have different weights—that usually big things weigh more than little ones, but that sometimes little things weigh more than big ones. All this he finds out experientially, and his knowledge is abstracted from the objects themselves. But I should like to give an example, just as primitive as that one, in which knowledge is abstracted from actions, from the coordination of actions, and not from objects. This example, one we have studied quite thoroughly with many children, was first suggested to me by a mathematician friend who quoted it as the point of departure of his interest in mathematics. When he was a small child, he was counting pebbles one day; he lined them up in a row, counted them from left to right, and got ten. Then, just for fun, he counted them from right to left to see what number he would get, and was astonished that he got ten again. He put the pebbles in a circle and counted them, and once again there were ten. He went around the circle in the other way and got ten again. And no matter how he put the pebbles down, when he counted them, the number came to ten. He discovered here what is known in mathematics as commutativity, that is, the sum is independent of the order. But how did he discover this? Is this commutativity a property of the pebbles? It is true that the pebbles, as it were, let him arrange them in various ways; he could not have done the same thing with drops of water. So in this sense there was a physical aspect to his knowledge. But the order was not in the pebbles; it was he, the subject, who put the pebbles in a line and then in a circle. Moreover, the sum was not in the pebbles themselves; it was he who united them. The knowledge that this future mathematician discovered that day was drawn, then, not from the physical properties of the pebbles, but from the actions that he carried out on the pebbles. This knowledge is what I call logical mathematical knowledge and not physical knowledge.

The first type of abstraction from objects I shall refer to as simple abstraction, but the second type I shall call reflective abstraction, using this term in a double sense. "Reflective" here has at least two meanings in the psychological field, in addition to the one it has in physics. In its physical sense reflection refers to such a phenomenon as the reflection of a beam of light off one surface onto another surface. In a first psychological sense abstraction is the transposition from one hierarchical level to another level (for instance, from the level of action to the level of operation). In a second psychological sense reflection refers to the mental process of reflection, that is, at the level of thought a reorganization takes place.

I should like now to make a distinction between two types of actions. On the one hand, there are individual actions such as throwing, pushing, touching, rubbing. It is these individual actions that give rise most of the time to abstraction from objects. This is the simple type of abstraction that I mentioned above. Reflective abstraction, however, is based not on individual actions but on coordinated actions. Actions can be coordinated in a number of different ways. They can be joined together, for instance; we can call this an additive coordination. Or they can succeed each other in a temporal order; we can call this an ordinal or a sequential coordination. There is a before and an after, for instance, in organizing actions to attain a goal when certain actions are essential as means to attainment for this goal. Another type of coordination among actions is setting up a correspondence between one action and another. A fourth form is the establishment of intersections among actions. Now all these forms of coordinations have parallels in logical structures, and it is such coordination at the level of action that seems to me to be the basis of logical structures as they develop later in thought. This, in fact, is our hypothesis: that the roots of logical thought are not to be found in language alone, even though language coordinations are important, but are to be found more generally in the coordination of actions, which are the basis of reflective abstraction. For the sake of completeness, we might add that naturally the distinction between individual actions and coordinated ones is only a gradual and not a sharply discontinuous one. Even pushing, touching, or rubbing has a simple type of organization of smaller subactions.

This is only the beginning of a regressive analysis that could go much further. In genetic epistemology, as in developmental psychology, too, there is never an absolute beginning. We can never get back to the point where we can say, "Here is the very beginning of logical structures." As soon as we start talking about the general coordination of actions, we are going to find ourselves, of course, going even further back into the area of biology. We immediately get into the realm of the coordinations within the nervous system and the neuron network, as discussed by McCulloch and Pitts. And then, if we look for the roots of the logic of the nervous system as discussed by these workers, we have to go back a step further. We find more basic organic coordinations. If we go further still into the realm of comparative biology, we find structures of inclusion ordering correspondence everywhere. I do not intend to go into biology; I just want to carry this regressive analysis back to its beginnings in psychology and to emphasize again that the formation of logical and mathematical structures in human thinking cannot be explained by language alone, but has its roots in the general coordination of actions.

$*$ 2 $*$

HAVING DEMONSTRATED that the roots of logical and
mathematical structures are to be found in the coordination
of actions, even before the development of language, I
should like now to look at how these coordinations of actions become mental operations, and how these operations
constitute structures. I shall start by defining what I mean
by an operation in terms of four fundamental characteristics.

First of all, an operation is an action that can be internalized; that is, it can be carried out in thought as well as
executed materially. Second, it is a reversible action; that is,
it can take place in one direction or in the opposite direction. This is not true of all actions. If I smoke my pipe
through to the end, I cannot reverse this action and have it
back again filled up with the same tobacco. I have to start
over again and fill it with new tobacco. On the other hand,
addition is an example of an operation. I can add one to one

and get two, and I can subtract one from two to get one again. Subtraction is simply the reversal of addition—exactly the same operation carried out in the other direction. There are types of reversibility that I should like to distinguish at this point. The first is reversibility by inversion or negation; for instance, $+A - A = 0$, or $+1 - 1 = 0$. The second is reversibility by reciprocity. This is not a negation, but is simply a reversal of order. For instance, $A = B$, the reciprocal is also true: $B = A$. The third characteristic of an operation is that it always supposes some conservation, some invariant. It is of course a transformation, since it is an action, but it is a transformation that does not transform everything at once, or else there would be no possibility of reversibility. For instance, in the case of arithmetical addition we can transform the way we group the parts together. We can say $5 + 1$, or $4 + 2$, or $3 + 3$, but the invariant is the sum. The fourth characteristic is that no operation exists alone. Every operation is related to a system of operations, or to a total structure as we call it. And I should like now to define what we mean by structure.

First of all, a structure is a totality; that is, it is a system governed by laws that apply to the system as such, and not only to one or another element in the system. The system of whole numbers is an example of a structure, since there are laws that apply to the series as such. Many different mathematical structures can be discovered in the series of whole numbers. One, for instance, is the additive group. The rules for associativity, commutativity, transitivity, and closure for addition all hold within the series of whole numbers. A second characteristic of these laws is that they are laws of transformation; they are not static characteristics. In the case of addition of whole numbers, we can transform one number into another by adding something to it. The third characteristic is that a structure is self-regulating; that is, in order to carry out these laws of transformation, we need not go outside the system to find some external element. Similarly, once a law of transformation has been applied, the result does not end up outside the system. Referring to the additive group once again, when we add one whole number to another, we do not have to go outside the series of whole numbers in search of any element that is not within the series. And once we have added the two whole numbers together, our result still remains within the series. We could call this closure, too, but it does not mean that a structure as a whole cannot relate to another structure or other structures as wholes. Any structure can be a substructure in a larger system. It is very easy to see that the whole numbers are a part of a larger system, which includes fractional numbers.*

I should like now to examine the three mother structures of the Bourbaki mathematicians and to raise the question of whether these mother structures correspond to anything

---

* The reader may ask here whether "structures" have real, objective existence or are only tools used by us to analyze reality. This problem is only a special case of a more general question: do relations have objective independent existence? Our answer will be that it is nearly impossible to understand and justify the validity of our knowledge without presupposing the existence of relations. But this answer implies that the word existence has to be taken to have a multiplicity of meanings.

natural and psychological or are straight and mathematical inventions established by axiomatization.*

As you know, the aim of the Bourbaki was to find structures that were isomorphic among all the various branches of mathematics. Up until that time, these branches, such as number theory, calculus, geometry, and topology, had all been more or less distinct and unrelated. What the Bourbaki set out to do was find forms or structures that were common to all these various contents. Their procedure was a sort of regressive analysis—starting from each structure in each branch and reducing it to its most elementary form. There was nothing a priori about it; it was the result of an inductive search and examination of mathematics as it existed. This search led to three independent structures that are not reducible one to the other. By making differentiations within each one of these structures or by combining two or more structures, all the others can be generated. For this reason the structures were called mother structures. Now the basic question for epistemology is whether these structures are natural in any sense as the natural numbers are, or whether they are totally artificial—simply the result of theorizing and axiomatizing. In an attempt to resolve this problem, let us look in more detail at each of the three mother structures.

The first is what the Bourbaki called the algebraic structure. The prototype of this structure is the mathematical notion of a group. There are all sorts of mathematical

---

* We shall not analyze the question here, but the more general concept of "category" already mentioned has equally a psychological counterpart. We refer the interested reader to Vol. XXIII of the *Etudes d'epistémologie génétique: Epistémologie et psychologie de la fonction* (1968).

groups: the group of displacements, as found in geometry; the additive group that I have already referred to in the series of whole numbers; and any number of others. Algebraic structures are characterized by their form of reversibility, which is inversion in the sense that I described above. This is expressed in the following way: $p \cdot p^{-1} = 0$, which is read as, "the operation p multiplied by the inverse operation p to the minus one equals zero."*

The second type of structure is the order structure. This structure applies to relationships, whereas the algebraic structure applies essentially to classes and numbers. The prototype of an order structure is the lattice, and the form of reversibility characteristic of order structures is reciprocity. We can find this reciprocity of the order relationship if we look at the logic of propositions, for example. In one structure within the logic of propositions, P and Q is the lower limit of a transformation, and P or Q is the upper limit. P and Q, the conjunction, precedes P or Q, the disjunction. But this whole relationship can be expressed in the reverse way. We can say that P or Q follows P and Q, just as easily as we can say that P and Q precedes P or Q. This is the form of reversibility that I have called reciprocity; it is not at all the same thing as inversion or negation. There is nothing negated here.

The third type of structure is the topological structure based on notions such as neighborhood, borders, and ap-

* The usual definition of algebraic structure as a set on which equivalence relations are defined leads to the same properties as the definition we use here (in particular: to every theory of the equivalence relations will correspond a theory of classes).

proaching limits. This applies not only to geometry but also to many other areas of mathematics. Now these three types of structure appear to be highly abstract. Nonetheless, in the thinking of children as young as 6 or 7 years of age we find structures resembling each of these three types, and I should like to discuss these here. Before I do, however, I shall tell a little story in an attempt to show that my drawing this parallel between the mother structures and children's operational structures is not completely arbitrary.

A number of years ago I attended a conference outside Paris entitled "Mental Structures and Mathematical Structures." This conference brought together psychologists and mathematicians for discussion of these problems. For my part, my ignorance of mathematics then was even greater than what I admit to today. On the other hand, the mathematician Dieudonne, who was representing the Bourbaki mathematicians, totally mistrusted anything that had to do with psychology. Dieudonne gave a talk in which he described the three mother structures. Then I gave a talk in which I described the structures that I had found in children's thinking, and to the great astonishment of us both we saw that there was a very direct relationship between these three mathematical structures and the three structures of children's operational thinking. We were, of course, impressed with each other, and Dieudonne went so far as to say to me: "This is the first time that I have taken psychology seriously. It may also be the last, but at any rate it's the first."

In children's thinking algebraic structures are to be found

quite generally, but most readily in the logic of classes—in the logic of classification. I shall take my example from the operations of simple classification, that is, just dividing a group of objects into piles according to their similarities, rather than the more complex procedure of multiplicative classification according to a number of different variables at the same time. Children are able to classify operationally, in the sense in which I defined that term earlier, around 7 or 8 years of age. But there are all sorts of more primitive classifying attempts in the preoperational stage. If we give 4- or 5-year-olds various cutout shapes—let's say simple geometric configurations like circles, squares, and triangles—they can put them into little collections on the basis of shape. The youngest children will make what I call figural collections; that is, they will make a little design with all the circles, and another little design with all the squares, and these designs will be an important part of the classification. They will think that the classification has been changed if the design is changed.

Slightly older children will forgo this figural aspect and be able to make little piles of the similar shapes. But while the child can carry out classifications of this sort, he is not able to understand the relationship of class inclusion. It is in this sense that his classifying ability is still preoperational. He may be able to compare subclasses among themselves quantitatively, but he cannot deduce, for instance, that the total class must necessarily be as big as, or bigger than, one of its constituent subclasses. A child of this age will agree that all ducks are birds and that not all birds are ducks. But then, if he is asked whether out in the woods there are more birds or more ducks, he will say, "I don't know; I've never counted them." It is the relationship of class inclusion that gives rise to the operational structure of classification, which is in fact analogous to the algebraic structures of the mathematicians. The structure of class inclusion takes the following form: ducks plus the other birds that are not ducks together form the class of all birds; birds plus the other animals that are not birds together form the class of all animals; etc. Or, in other terms, $A + A' = B$, $B + B' = C$, etc. And it is easy to see that this relationship can readily be inverted. The birds are what is left when from all the animals we subtract all the animals but the birds. This is the reversibility by negation that we mentioned earlier: $A - A = 0$. This is not exactly a group; there is inversion, as we have seen, but there is also the tautology, $A + A = A$. Birds plus more birds equal birds. This means that distributivity does not hold within this structure. If we write $A + A - A$, where we put the parentheses makes a difference in the result. $(A + A) - A = 0$, whereas $A + (A - A) = A$. So it is not a complete group; it is what I call a grouping, and it is an algebra-like structure.

Similarly, there is a very primitive ordering structure in children's thinking, just as primitive as the classification structure. A very simple example is the structure of seriation. We have given children the following problem. First we present them with a collection of sticks of different lengths. The differences in length are small enough so that it takes a careful comparison to detect them; this is not an easy perceptual task. Some are between ⅛ and a ¼ inch different in length, and there are about ten such sticks, the smallest being about 2 inches long. Then we ask the children to put them in order from the smallest to the biggest. Preoperational children approach this problem without any structural framework (in the sense that I have been describing structures). That is, they take a big one and a little one, and then another big one and a little one, and then another big one and a little one, but they make no coordinations among these pairs of sticks; or they may take three at a time—a little one, a middle-sized one, and a big one—and make several trios. But they will not manage to coordinate all the sticks together in a single series. Slightly older children at the end of the preoperational stage succeed in putting all the sticks together in a series, but only by trial and error; they do not have any systematic approach. By contrast, children from about the age of 7 years have a totally different way of going about this problem. It is a very exhaustive systematic approach. They first of all find the very smallest stick, then they look through the remaining sticks for the smallest ones left, then they look for the smallest one that is left again, and so on until the whole structure, the whole series, has been built. The reversibility implied here is one of reciprocity. When the child looks for the smallest stick of all those that remain, he understands at one and the same time that this stick is bigger than all the ones he has taken so far and smaller than all the ones that he will take later. He is coordinating here at the same time the relationship "bigger than" and the relationship "smaller than."

There is even more convincing evidence of the operational nature of this structure, and that is the fact that at the same time children become capable of reasoning on the basis of transitivity. Let us say that we present two sticks to a child, stick A being smaller than stick B. Then we hide stick A and show him stick B together with a larger stick C. Then we ask him how A and C compare. Preoperational children will say that they do not know because they have not seen them together—they have not been able to compare them. On the other hand, operational children, the children who proceed systematically in the seriation of the sticks, for instance, will say right away that C is bigger than A, since C is bigger than B and B is bigger than A. According to logicians, seriation is a collection of asymmetrical, transitive relationships. Here we see quite clearly that the asymmetrical relationships and the transitivity do indeed develop hand in hand in the thinking of small children. It is very obvious, moreover, that the structure here is one whose reversibility is reciprocity and not negation. The reversibility is of the following sort: A is smaller than B implies that B is larger than A, and this is not a negation but simply a reciprocal relationship.

The third type of structure, according to the Bourbaki mathematicians, is the topological structure. The question of its presence in children's thinking is related to a very interesting problem. In the history of the development of geometry, the first formal type was the Euclidian metric geometry of the early Greeks. Next in the development was projective geometry, which was suggested by the Greeks but not fully developed until the seventeenth century. Much later still, came topological geometry, developed in the nineteenth century. On the other hand, when we look at the theoretical relationships among these three types of geometry, we find that the most primitive type is topology and that both Euclidian and projective can be derived from topological geometry. In other words, topology is the common source for the other two types of geometry. It is an interesting question, then, whether in the development of thinking in children geometry follows the historic order or the theoretical order. More precisely, will we find that Euclidian intuitions and operations develop first, and topological intuitions and operations later? Or will we find that the relationship is the other way around? What we do find, in fact, is that the first intuitions are topological. The first operations, too, are those of dividing space, of ordering in space, which are much more similar to topological operations than to Euclidian or metric ones.

I should like to give you a couple of examples of the topological intuitions that exist at the preoperational level. Preoperational children can of course distinguish various Euclidian shapes—circles from rectangles, from triangles, etc.—as Binet has shown. They can do this at about 4 years of age, according to his norms. But let us look at what they do before this age. If we show them a circle and ask them to copy it in their own drawing, they will draw a more or less circular closed form. If we show them a square and ask them to copy it, they will again draw a more or less circular closed form. Once again, if we show them a triangle, they will draw just about the same thing. Their drawings of these shapes are virtually indistinguishable. But if, on the other hand, we ask them to draw a cross, to copy a cross, they will draw something totally different from their drawings of the closed figures. They will draw an open figure, two lines that more or less come to a cross or touch each other. In general, then, in these drawings we see that the children have not maintained the Euclidian distinctions in terms of different Euclidian shapes, but that they have maintained the topological distinctions. Closed shapes have been drawn as closed, and open shapes have been drawn as open.

Perceptually, of course, children do recognize distinctions among Euclidian shapes, but in their representations of these shapes to themselves they seem not to make such distinctions. One might think that this is just a question of muscle control, that the children are not able to draw squares. But we can give them another problem that demands seemingly just as much muscle control. We can show them three different figures in which there is a large circle and a small circle, but in the first the small circle is inside the larger, in the second the small circle is outside the larger, and in the third the small circle is on the border— half inside, half outside. Three-year-olds who do not yet draw squares as distinct from circles nonetheless copy these figures accurately, at least preserving the relationships of inside, outside, and on the border. Some children even find descriptive ways of referring to the third figure, saying that the small circle is half outside, for instance. This implies that they see it as not inside and not outside but on the border, and all these are topological relationships.

Some authors have maintained that the distinction between rectilinear and curvilinear figures is just as primitive as these distinctions among inside, outside, and on the border. Rectilinear and curvilinear figures, of course, have no distinction within topology; they are only different within Euclidian geometry. In reply to these authors, I should like to cite the work of two Montreal psychologists, Monique Leurendau and Adrien Pinard. These psychologists repeated all our research on geometry and spatial representation, taking twenty subjects at each age and doing every experiment with each one of the subjects, which is something that we have never done. And they proceeded to do a very thorough analysis, both qualitatively and statistically, of the behavior of each of these children. They used ordinal statistics, such as Gutman developed. Their analysis revealed that, indeed, sometimes children seemed to be distinguishing curvilinear from rectilinear figures, but in every instance they were actually using topological relationships to make the distinction. That is, the figures were different in topological relationships as well as in the Euclidian relationships of straight lines or curved lines, and the children were basing their judgments on the topological aspects of the figures.

So far I have attempted to demonstrate that the three mathematical mother structures have natural roots in the development of thinking in individuals. I should like now to show how, in children's thinking, other structures can develop out of combinations of two or more of the basic structures. I indicated earlier that this is the source of the many and varied mathematical structures in all the different branches of mathematics. The example I shall take from psychology is the notion of number, which is not based on only one of the three primitive structures, but rests on a combination of two of them.

I have referred to the operation used by Cantor in the construction of transfinite numbers, namely, the operation of one-to-one correspondence. Let us start now by looking at how this operation develops in children's thinking. We have done an experiment of the following sort. We line up, let us say, eight red tokens in front of a child and then give him a pile of blue tokens and ask him to put out just as many blue tokens as there are red ones. In a very early stage, the child will make a line of blue tokens about as long as the line of red tokens but will pay no attention to whether or not there is actually the same number of blues as reds. A little more sophisticated behavior is to operate on the basis of one-to-one correspondence, that is, taking a blue token and putting it right underneath a red one. But this is what I call optical correspondence, because the child will consider that the one-to-one correspondence depends upon this tight spatial relationship between each blue and each red. If we change the spatial disposition without adding, or taking away any tokens—we simply spread out or squeeze up one of the lines—the child will say that things are changed now and that there are no longer as many blue tokens as red ones. If we count one row and get eight and then ask him how many tokens he thinks are in the other

What are the questions asked? E.g. "Are there more?" Does the child know the word "more"? Is the problem linguistic?

row, which has been spread out, he will say, "There must be nine or ten." Even if he counts each row, eight in the shorter row and eight in the longer row, he will say, "Yes, there are eight here and eight there, but still there are more there; it is longer." Finally, the one-to-one correspondence becomes operational, and at that time there is conservation of number in the sense of the realization that the number does not change just because the spatial arrangement changes. In this instance, once the child has established one-to-one correspondence by taking a blue token for every red one, no matter how we change the shapes, he will be able, without counting or even without thinking very hard, to say that the numbers must still be the same because of the one to one correspondence that he established at the outset. One-to-one correspondence seems to be, then, the basis for the notion of number.

This brings to mind immediately Russell and White-head's work in *Principia Mathematica*, where they define a number as the class of equivalent classes—equivalent in the sense of numerical equivalence established through one-to-one correspondence. If we have a class that consists of five people, for instance, and a class that consists of five trees, and a class that consists of five apples, what these three classes have in common is the number 5. And it is in this sense that Russell and Whitehead state that a number is a class of equivalent classes. Now this view of the basis for the idea of a number does seem to be justified, as I said a moment ago, since, in fact, the number seems to be derived from one-to-one correspondence. But there are two types of one-to-one correspondence, and it is important for us to look at which type Russell and Whitehead used.

On the one hand, there is one-to-one correspondence based on the qualities of the elements. An element of one class is made to correspond to a specific element of another class because of some qualities that the two classes have in common. Let us suppose, for instance, that the classes we mentioned a moment ago (five people, five trees, five apples) are paper cutouts and that five different colors of paper are used. Therefore there are five paper people—red, orange, green, yellow, and blue; five paper trees, one of each of the same colors; and five paper apples, again of the same colors. The qualitative one-to-one correspondence would consist of putting the red person in correspondence with the red tree and the red apple, the green person in correspondence with the green tree and the green apple, etc. This is, in fact, the procedure of double classification—constructing a matrix by classifying on two dimensions at once.

The other type of one-to-one correspondence is not based on the qualities of the individual elements. Russell and Whitehead's famous example of equivalent classes makes a correspondence between the months of the year, Napoleon's marshals, the twelve apostles, and the signs of the zodiac. In this example there are no qualities of the individual members that lead to a specific correspondence between one element of one class and one element of another. We cannot say, for instance, that St. Peter corresponds to the month of January, which corresponds to Marshal Ney, who corresponds to Cancer. When we say that these four groups correspond to one another, we are using one-to-one correspondence in the sense that any element can be made to correspond to any other element. Each element counts as one, and its particular qualities have no importance. Each element becomes simply a unity, an arithmetic unity.

Now this is a very different operation from the operation of one-to-one correspondence based on qualities, which is used in classification and which gives rise to matrices, as I just described. The one-to-one correspondence, in which any element can correspond to any other element, is a very different operation. Elements are stripped of their qualities and become arithmetic unities. Now it is very clear that Russell and Whitehead have not used the qualified one-to-one correspondence that is used in classification. They have used the correspondence in which the elements become unities. They are, therefore, not basing number only on classification operations as they intend. They have, in fact, got themselves into a vicious circle, because they are attempting to build the notion of number on the basis of one-to-one correspondence, but in order to establish this correspondence they have been obliged to call upon an arithmetic unity, that is, to introduce a notion of a non-qualified element and numerical unity in order to carry out the one-to-one correspondence. In order to construct numbers from classes, they have introduced numbers into classes.

Their solution, then, does not turn out to be an adequate one. The problem of the basis of the notion of number, the epistemological problem, remains, and we must look for another solution. Psychological research seems to offer one. When we study the development of the notion of number in children's thinking, we find that it is not based on classifying operations alone but that it is a synthesis of two different structures. We find that along with the classifying structures, which are an instance of the Bourbaki algebraic structures, number is also based on ordering structures, that is, a synthesis of these two different types of structures. It is certainly true that classification is involved in the notion of number. Class inclusion is involved in the sense that two is included in three, three is included in four, etc. But we also need order relationships, for this reason: if we consider the elements of the classes to be equivalent (and this of course is the basis of the notion of number), then by this very fact it is impossible to distinguish one element from another—it is impossible to tell the elements apart. We get the tautology $A + A = A$; we have a logical tautology instead of a numerical series. Given all these elements, then, whose distinctive qualities we are ignoring, how are we going to distinguish among them? The only possible way is to introduce some order. The elements are arranged one after another in space, for instance, or they are considered one after another in time, or they are counted one after another. This relationship of order is the only way in which elements, which are otherwise being considered as identical, can be distinguished from one another.

In conclusion, then, number is a synthesis of class inclusion and relationships of order. It depends on an algebraic type of structure and an ordering type of structure, both at one time. One type of structure alone is not adequate.

I think that it is really quite obvious, if not trite, that number is based on two different types of operation. In fact, if we look at any theoretical formulation of number, we will find that in the number theories based on ordination there is always an element of inclusion. Similarly, in theories based on cardination there is always an element of order.

I should like to discuss one final area before leaving this analysis of the types of operational structures used in children's logical thinking. At the level of concrete operations being examined, that is, from the age of 6 or 7 years to the age of 11 or 12, there are two types of reversibility: negation and reciprocity. But they are never synthesized in a single system, so that it is possible to go from one type of reversibility to the other within the same system. At the level of formal operations, which, as I have said, start to appear at about 11 or 12 years of age, new logical structures are built that give rise, for instance, to the logic of propositions in which both types of reversibility are equally available. For example, we can look at this implication: P implies Q; its negation is P and not Q. But the reciprocal, Q implies P, is just as readily available within the system, and it too has its negation, Q and not P. This last has a new relationship with respect to the initial implication, and we can call it the correlative.

This more complex type of structure is brought into evidence when we give children problems involving double frames of reference and space—for instance, problems of relative motion. Let us say that we have a snail on a little board. If the snail moves to the right, we can take that as the direct operation. And the inversion, or negation, would be the snail moving to the left. But the reciprocal of a move by the snail to the right would be a move by the board to the left, and then the correlative would be a move by the board to the right. If the snail moves to the right on the board and at the same time the board moves to the left, with respect to an external frame of reference, it is as if the snail did not move at all, with respect to an external frame of reference. With respect to an external frame, there are two ways of reversing the snail's motion: one is for the snail to move back again; the other is for the board to move. Before children are able to synthesize the two types of reversibility in the single system, that is, before the age of 11 or 12 years, they cannot resolve problems of this sort, which require a coordination between two different types of motion with two possible frames of reference.

What is the experiment with kids?

$*$ 3 $*$

I HAVE DISCUSSED the logical mathematical structures. Now I should like to write briefly about the relationship between these structures and language on the one hand, and the relationship between these structures and sensory-motor activities on the other hand, in order to deal with the problem that I raised. The decisive argument against the position that logical mathematical structures are derived uniquely from linguistic forms is that, in the course of intellectual development in any given individual, logical mathematical structures exist before the appearance of language. Language appears somewhere about the middle of the second year, but before this, about the end of the first year or the beginning of the second year, there is a sensory-motor intelligence that is a practical intelligence having its own logic—a logic of action. The actions that form sensory-motor intelligence are capable of being repeated and of being generalized. For example, a child who has learned to pull a blanket toward him in order to reach a toy that is on it then is capable of pulling the blanket to reach anything else that may be placed on it. The action can also be generalized so that he learns to pull a string to reach what is attached to the end of the string, or so that he can use a stick to move a distant object.

Whatever is repeatable and generalizable in an action is what I have called a scheme, and I maintain that there is a logic of schemes. Any given scheme in itself does not have a logical component, but schemes can be coordinated with one another, thus implying the general coordination of actions. These coordinations form a logic of actions that are the point of departure for the logical mathematical structures. For example, a scheme can consist of subschemes or subsystems. If I move a stick to move an object, there is within that scheme one subscheme of the relationship between the hand and the stick, a second subscheme of the relationship between the stick and the object, a third subscheme of the relationship between the object and its position in space, etc. This is the beginning of the relationship of inclusion. The subschemes are included within the total scheme, just as in the logical mathematical structure of classification subclasses are included within the total class. At the later stage this relationship of class inclusion gives rise to concepts. At the sensory-motor stage a scheme is a sort of practical concept.

Another type of logic involved in the coordination of schemes is the logic of order: for instance, in order to achieve an end we have to go through certain means. In this example there is an order between the means and the goal. And, once again, it is practical order relationships of this sort that are the basis of the later logical mathematical structures of order. There is also a primitive type of one-to-one correspondence. For instance, when an infant imitates a model, there is a correspondence between the model on the one hand and his imitation on the other. Even when he imitates himself, that is, when he repeats an action, there is a correspondence between the action as carried out one time and the action as carried out the next.

In other words, we find here, in sensory-motor intelligence, a certain logic of inclusion, a certain logic of ordering, and a certain logic of correspondence, which I maintain are the foundations for the logical mathematical structures. They are certainly not operations, but they are the beginnings of what will later become operations. We can also find in this sensory motor intelligence the beginnings of two essential characteristics of operations, namely, a form of conservation and a form of reversibility.

The conservation characteristic of sensory-motor intelligence takes the form of the notion of the permanence of an object. This notion does not exist until near the end of the infant's first year. If a 7- or 8-month-old is reaching for an object that is interesting to him and we suddenly put a screen between the object and him, he will act as if the object not only has disappeared but also is no longer accessible. He will withdraw his hand and make no attempt to push aside the screen, even if it is as delicate a screen as a handkerchief. Near the end of the first year, however, he will push the screen aside and continue to reach for the object. He will even be able to keep track of a number of successive changes of position. If the object is put in a box and the box is put behind a chair, for instance, the child will be able to follow these successive changes of position. This notion of the permanence of an object, then, is the sensory-motor equivalent of the notions of conservation that develop later at the operational level.

Similarly we can see the beginnings of reversibility in the understanding of spatial positions and changes of position, that is, in the understanding of movement in space within which the child moves at the time of the culmination of sensory-motor intelligence. At the beginning of the second year children have a practical notion of space which includes what geometers call the group of displacements, that is, the understanding that a movement in one direction can be canceled by a movement in the other direction—that one point in space can be reached by one of a number of different routes. This of course is the detour behavior that psychologists have studied in such detail in chimpanzees and infants.

So this is again practical intelligence. It is not at the level of thought, and it is not at all in the child's representation, but he can act in space with this amount of intelligence. Furthermore, this kind of organization exists as early as the second half of the first year before any use of language for expression. And this is my first argument.

My second argument deals with children whose thinking

is logical but who do not have language available to them—namely, the population of the deaf and dumb. Before I discuss the experimental findings on intelligence in deaf and dumb children, I should like to discuss briefly the nature of representation. Between the age of about 1½ years and the age of 7 or 8 years when the operations appear, the practical logic of sensory-motor intelligence goes through a period of being internalized, of taking shape in thought at the level of representation rather than taking place only in the actual carrying out of actions. I should like to insist here upon one point that is too often forgotten: that there are many different forms of representation. Actions can be represented in a number of different ways, of which language is only one. Language is certainly not the exclusive means of representation. It is only one aspect of the very general function that Head has called the symbolic function. I prefer to use the linguists' term: the semiotic function. This function is the ability to represent something by a sign or a symbol or another object. In addition to language the semiotic function includes gestures, either idiosyncratic or, as in the case of the deaf and dumb language, systematized. It includes deferred imitation, that is, imitation that takes place when the model is no longer present. It includes drawing, painting, modeling. It includes mental imagery, which I have characterized elsewhere as internalized imitation. In all these cases there is a signifier which represents that which is signified, and all these ways are used by individual children in their passage from intelligence that is acted out to intelligence that is thought. Language is but one among these many aspects of the semiotic function, even though it is in most instances the most important.

This position is confirmed by the fact that in deaf and dumb children we find thought without language and logical structures without language. Oleron in France has done interesting work in this area. In the United States I should like to mention especially the work of Hans Furth and his excellent book, *Thinking without Language*. Furth finds a certain delay in the development of logical structures in deaf and dumb children as compared with normal children. This is not surprising since the social stimulation of the former is so limited, but apart from this delay the development of the logical structures is similar. He finds classifications of the sort discussed before; he finds seriation of the type discussed before; he finds correspondence; he finds numerical quantity; and he finds the representation of space. In other words, there is well-developed logical thinking in these children even without language.

Another interesting point is that, although deaf and dumb children are delayed with respect to normal children, they are delayed much less than children who have been blind from birth. Blind infants have the great disadvantage of not being able to make the same coordinations in space that normal children are capable of during the first year or two, so that the development of sensory-motor intelligence and the coordination of actions at this level are seriously impeded in blind children. For this reason we find that there are even greater delays in their development at the level of representational thinking and that language is not sufficient to compensate for the deficiency in the coordination of actions. The delay is made up ultimately, of course, but it is significant and much more considerable than the delay in the development of logic in deaf and dumb children.

In approaching my third argument I should like to point out again that Chomsky has reversed the position of the logical positivists on the question of the relationship between logic and language. According to Chomsky, logic is not derived from language, but language is based on a kernel of reason. Transformational grammars, in whose development Chomsky played a leading role, seem to me to be of great interest and to show very clear similarities to the operations of intelligence that have been discussed. Chomsky goes so far as to say that the kernel of reason on which the grammar of language is constructed is innate, that it is not constructed through the actions of the infant as I have described but is hereditary and innate. I think that this hypothesis is unnecessary, to say the least. In point of fact, it is very striking that language does not appear in children until the sensory-motor intelligence is more or less achieved. I agree that the structures that are available to a child at the age of fourteen or sixteen months are the intellectual basis upon which language can develop, but I deny that these structures are innate. I think that we have been able to see that they are the result of development. Hence the hypothesis that they are innate is, as I have said, unnecessary. The main thing that I should like to emphasize in Chomsky's position is that he has reversed the classical view that logic is derived from language by maintaining that language is based on intellectual structures.

My final argument will be based on the work of Madame Hermine Sinclair, who studies the relationships between operational level and linguistic level in children between 5 and 8 years of age. Mme Sinclair was a linguist before she came to study psychology in Geneva, and at her first contact with our work she was convinced that the operational level of children simply reflected their linguistic level; that is, she was maintaining the position of the logical positivists. I suggested to her that she study this question, since it had never been investigated closely, and see what relationship existed between the operational level and the linguistic level of children. As a result Mme Sinclair performed the following experiment. First she established two groups of children. One group consisted of conservers; that is, they realized that, when a certain amount of liquid was poured from a glass of one shape into a glass of another shape, the quantity did not change in spite of the appearances. The other group consisted of nonconservers: they judged the quantity of liquid according to its appearance and not according to any correlation between height and width, or any reasoning in terms of the fact that no liquid had been added or taken away. Then Mme Sinclair proceeded to study the language of each of these groups of children by giving them very simple objects to describe. Usually she presented the objects in pairs, so that the children could describe them by comparing them, or could describe each object by itself. She gave them, for instance, pencils of different widths and lengths. She found noticeable differences in the language used to describe these objects according to whether the child was a conserver or a nonconserver. Nonconservers tended to describe objects in terms that linguists call scalers. That is, they would describe one object at a time and one characteristic at a time—"That pencil is long"; "That pencil is fat"; "It is short"; and similar observations. The conservers, on the other hand, used what linguists call vectors. They would keep in mind both the objects at once and more than one characteristic at once. They would say, "This pencil is longer than that one, but that one is fatter than this one" –sentences of that sort.

So far the experiment seems to show a relationship between operational level and linguistic level. But we do not yet know in what sense the influence is exercised. Is the linguistic level influencing the operational level, or is the operational level influencing the linguistic progress? To find the answer, Mme Sinclair went on to another aspect of this experiment. She undertook to give linguistic training to the nonconserving group. Through classical learning theory methods she taught these children to describe the objects in the same terms that the conservers used. Then she examined again the children who had previously been nonconservers but who had then learned the more advanced linguistic forms to see whether this training had affected their operational level. (Let me point out that she did this experiment in several different areas of operations, not only for conservation but also for seriation and other areas.) Well, in every case she found that there was only minimal progress after the linguistic training. Only 10 per cent of the children advanced from one substage to another. This is such a small proportion that it leads one to wonder whether these children were not already at an intermediate phase and right on the threshold of the next substage. Mme Sinclair's conclusion on the basis of these experiments is that the intellectual operations appear to give rise to linguistic progress, and not vice versa.

I should like to leave this discussion of language and logic now and look at the type of thinking, the type of logical reasoning, that children are capable of in what I call the preoperational stage, that is, ages 4, 5, and 6 years, before the onset or the development of logical operations. Although logical structures are not fully developed at the preoperational stage, we do find what can be called semilogic. In my earlier works I used to call this articulated intuitions, but since then we have done a good deal more work in this area. It seems quite clear now that the thought of children of these ages is characterized by semilogic in a very literal sense; it is a half logic. We have here operations that are lacking in reversibility; they work only in one direction. This logic, then, consists of functions in the mathematical sense of that term, that is, as described by mathematicians: $y = (f)x$. A function in this sense represents an ordered couple or an application, but an application that moves always in one direction. This kind of thinking leads to the discovery of dependency relationships and of covariations, that is, that variations in one object are correlated with variations in another.

The remarkable thing about these functions is that they

do not lead to conservation. Here is one example. We have a piece of string attached to a small spring. It extends out horizontally, goes around a pivot, and hangs down vertically. Now when we put a weight on the end of the string, or increase a weight already there, the string is pulled so that the part that is hanging down vertically is lengthened with respect to the part that is horizontal. Five-year-olds are perfectly able to grasp the relationship that with the greater weight the vertical part is longer and the horizontal part is shorter, and that when the vertical part becomes shorter the horizontal part becomes longer. But this does not lead to conservation. The sum of the vertical part and the horizontal part does not stay the same for these children.

Here is another example of a function in the sense of an application. We give children a number of cards, on each of which there is a white part and a red part, and also give them a number of cutouts of different shapes. Their task is to find a cutout that will cover up the red part on the card. It need not correspond exactly, but it simply must cover the red part completely. The interesting thing is that these children understand the relationship many-to-one, since they realize that there are a number of different cutout shapes all of which can completely cover the red, but this does not permit them to construct a good classification system based on the relationship of one-to-many. Here is another case of half of a logical structure. In the language of the Bourbaki mathematicians many-to-one is a function, but one-to-many is not a function.

More generally, the reason that functions are so interesting is that they demonstrate still more clearly the importance of relationships of order in preoperational thinking. A great many relationships that for us are metric are simply ordinal for children: measure does not come into their judgments at all.* A very good example is the conservation of length, which has been mentioned. If two sticks are the same length when they are side by side, and one is pushed over to the side, we continue to judge them to be the same length because we take into account both ends, and we realize that the important thing is the distance between the left-hand end and the right-hand end in each case. Preoperational children, however, do not base their judgments on the order of the end points. If they are looking at one end of the sticks, their judgment of length is based on which one goes farther in that direction. There are a great many experiments in which children's reactions are based on ordinal relationships rather than on quantitative ones, and it seems to me that this occurs because they are using a logic of functions rather than a complete operational logic.

Another characteristic of this semilogic is the notion of identity, which precedes the notion of conservation. We have seen that there is a certain notion of identity in sensory-motor intelligence, and a child realizes that an object has a

* I am aware of the fact that not all logicians follow the Bourbaki school, and that for intuitionists the intuition of the series of numbers is more fundamental than the concept of set or of structure. This corresponds psychologically to the fact that purely ordinal tasks are sometimes transformed by the child into quantitative ones. It will be a task for the future to analyze the relation between these two types of logic and these two types of behavior.

certain permanency. This is not a case of conservation in the sense that we use the term, since the object does not change its form in any way—it simply changes its position. But it is a case of identity, which is one of the starting points for the later notion of conservation. We have also studied the notion of identity in the preoperational thinking of children from the age of about 4 years. We have found that there is nothing more variable than the notion of identity, which by no means remains the same throughout the intellectual development of a child. What it means for something to preserve its identity changes according to the age of the child and according to the situation in which the problem is presented.

The first thing to keep in mind is that identity is a qualitative notion and not a quantitative one. For instance, a preoperational child who maintains that the quantity of water changes according to the shape of its container will nonetheless affirm that the water is the same—only the quantity has changed. My colleague, Jerome Bruner, thinks that a notion of the principle of identity is sufficient as a foundation for the notion of conservation. For my part, I find this position questionable. To have the principle of identity one has only to distinguish between that which changes in a given transformation and that which does not change. In the case of the pouring of liquids, children have only to make a distinction between the form and the substance. But in the notion of conservation, more is required. Quantification is rather more complex, as we have seen, especially since the most primitive quantitative notions are the ordinal notions just described, which are not adequate in all cases of quantitative comparison. It is not until children also develop the operations of compensation and reversibility that the quantitative notion of conservation is established.

But I should like to give some new examples of how the notion of identity changes with development. We have done a number of different experiments, in which Gilbert Boyat has been one of the principal collaborators. In this research we have found a first level where identity is semi-individual and semigeneric. A child will believe that objects are identical to the extent that one can do the same things with these objects. For example, a collection of beads on a table is recognized as being the same as the same beads in the form of a necklace, because one can take the necklace apart and make a pile of the beads or string them together again and make a necklace. Or a piece of wire in the shape of an arc is recognized as being the same piece of wire as when it is straight, because it can be bent into an arc or straightened into a straight line. A little later the child becomes slightly more demanding in his criteria for identity. It is no longer sufficient that an object be assimilated to a certain scheme. The identity becomes more individualized. At this stage he will say that a piece of wire is no longer the same piece when it is in the shape of an arc, because it no longer has the same form.

One interesting experiment of this sort came up rather fortuitously in the course of another experiment. Children were ordering squares according to size, and in the course of this activity one child put a square on a corner instead of along the edge and then rejected it, saying that it was no longer a square. We then started another experiment in which we investigated this more closely, presenting a cutout square in different positions and asking questions of the following types. Is it the same square? Is it still a square? Is it the same piece of cardboard? Are the sides still the same length? Are the diagonals still the same length? We put these questions, of course, in terms that made sense to the children we were interviewing. We found that the children, until the age of about 7 years, denied the identity: it is no longer a square; it is no longer the same square; the sides are no longer the same length; it is longer now in this direction; the angles are no longer right angles; etc.

In the area of perception there are similar experiments to be done. We are all familiar with the phenomenon of apparent motion or stroboscopic motion. One object appears and disappears, and as it disappears another object appears, and as the second object disappears the first object appears again. If this is done at the right speed, it looks as if one and the same object is moving back and forth between the two positions. It occurred to me that it would be interesting to study identity through this phenomenon of stroboscopic motion, by having one of the objects a circle and the other a square. When the object moves to one side it looks as if it is becoming a circle, and when it moves to the other side it looks as if it is becoming a square. It looks like a single object that is changing its shape as it changes its position. First of all, I should point out that it is much easier for children to see this apparent motion than for adults. The thresholds are much wider. Almost any speed of alternation or a great range of speeds of alternation gives rise to this impression of apparent motion for children, whereas for adults the limits are much narrower. The interesting thing in this experiment of ours is that, in spite of this facility that children have in seeing stroboscopic motion, they tend to deny the identity of the object. They will say, "It is a circle until it gets almost over to here, and then it becomes a square," or "It is no longer the same object—one takes the place of the other." Adults, on the other hand, see a circle that turns into a square and a square that turns into a circle. They find this strange, but nonetheless it is what they see: the same object changing its shape. The notion of identity, then, in this experiment rises very clearly as a function of age. And this is only one of many experiments in which we have found similar results.

The last experiment that I should like to mention is carried out by Boyat on plant growth. He started by experimenting with the growth of a bean plant, but that took too long, so instead he uses a chemical in a solution, which grows in a few minutes into an arborescent shape looking something like a seaweed. Periodically, as a child watches this plant grow, he is asked to draw it, and then with his drawings as reminders he is asked whether, at the various points in its growth, it is still the same plant. We refer to the plant by the same term the child uses for it—a plant, seaweed, macaroni—whatever he happens to use. Then we ask him to draw himself when he was a baby, and himself a little bigger, and still a littler bigger, and as he is now. And we ask the same questions as to whether all these drawings are drawings of the same person, whether the person is always he. At a relatively young age, a child will deny that the same plant is represented in his various drawings. He will say that this is a little plant, and that is a big plant—it is not the same plant. In referring to the drawings of himself, however, he will be likely to say that all show the same person. Then, if we go back to the drawings of the plant, some children will be influenced by their thoughts about their drawings of themselves and will say now that they realize that it is the same plant in all the drawings, but others will continue to deny this, maintaining that the plant has changed too much, that it is a different plant now. Here, then, is an amusing experiment which shows that the changes taking place within the logical thinking of children as they grow older affect even the notion of identity itself. Even identity changes in this field of continual transformation and change.*

* Philosophers have often asked under what conditions "things" or "persons" remain the same. We want to stress that strict identity (in the logical or Leibnizian sense) cannot be meant in these controversies. Strict identity corresponds to the semantical fact that one concept or one object may in a given language have several names. This being taken into account, it is obvious that our experiments refer often to physical identity or psychological identity, relations the evolution of which we have been able to follow, and which are characterized by the fact that in contradistinction with the concepts of number, space, tune they do not reach, even for the adult, a stage of stable equilibrium. This fact, illuminated by our genetic analysis, may explain the heated controversies about physical or psychological identity in the Anglo-Saxon literature.

* 4 *

Let us look more closely now at the development of the notions of speed and time. The traditional view of speed and time leads us into a vicious circle. Speed is defined as a relationship between time and space, and yet time can be measured only on the basis of a constant speed. This sets the stage for a study in genetic epistemology, namely, the search to find out whether one of these two notions is more basic than the other, and whether we can escape from the vicious circle by deriving the less basic notion from the more primitive one. The hypothesis that I am going to defend here is that the more primitive notion is the more complex and the less differentiated, namely, the notion of movement, including speed. I shall try to show that time can be defined as a coordination of movements or of speeds in the same sense as space is a coordination of changes of position. Changes of position, of course, are simply movements considered without taking the speed of the movement into account. Space, then, is a coordination of movements without taking speeds into account, whereas time, in my hypothesis, is the coordination of movements, including their speeds.

We come here to the striking parallel that exists between time and space. This is a classical parallel, of course, found throughout the writings of Newton, of Kant, and of an endless number of other philosophers, to relativity theory, in which the two are partially fused. But there are three important differences between space and time in spite of this parallelism, and I should like to discuss them. First of all, time is irreversible; unfortunately, once we have lived through a day we cannot go back and live through it again. Movements in space, however, are reversible; we can go from point A to point B, and then we can go back from point B to point A. The second difference is that space can be considered separately from its contents. It is true that one aspect of space is tied to its contents and cannot be separated from it, namely, physical space, as in relativity theory. Nonetheless, we can consider space separately from its contents. The science of this independent space is the science of pure geometry—pure in the sense that it is in no way limited by physical space. Time, on the other hand, cannot be considered separately from its contents. Time is always tied to speeds, and speeds invariably have not only a physical but also a psychological reality. We cannot create a pure science of time or a pure chronometry in the same way that we can create a pure geometry. The third difference, which is of great importance psychologically, is that we can perceive a whole geometric figure. Let us consider a figure as simple as a straight line—we can perceive a whole line as simultaneous. A temporal duration, however, no matter how short it is, cannot be apprehended all at once. Once we are at the end of it, the beginning can no longer be perceived. In other words, any knowledge of time presupposes a reconstruction on the part of the knower, since the beginning of any duration has already been lost and we cannot go back in time to find it again. Knowledge of space is therefore much more direct and simpler from the psychological point of view than knowledge of time.

I should like now to develop my hypothesis that the notion of speed is more fundamental than the notion of time, and that time is a coordination of speeds, by examining what is involved in both the notion and the perception of speed. But before I do that, I must make clear one distinction that will be important in what I should like to say: when we are considering temporal notions, there are two different kinds. The first is the notion of temporal order or the succession of events (A comes before B, B comes before C, C comes before D, etc.). The second is the interval between two events, that is, the length of time from A to B, and the length of time from B to C. It is quite clear that the order of temporal events can be considered without paying any attention to the duration or the interval of time. We shall use the term duration to refer to the intervals between temporal events, and the term order to refer to the simple succession of events when we are not paying any attention to the time intervals.

We have found that the classical notion of speed as a relationship between the spatial interval and the temporal duration appears very late in child development, about 9 or 10 years of age. By contrast, as early as the preoperational period, that is, even before the age of 6 years, there are intuitions of speed that are not based on this ratio. This primitive intuition is based on succession; it is an ordinal intuition and is not based on durations. This notion of speed that is not based on temporal duration turns out to be important in our attempt to escape from the vicious circle. This early intuition is based on the phenomenon of passing. If one moving object catches up with and overtakes another moving object, even very small children will say that the former object is going faster than the latter. This primitive intuition of speed* based on overtaking is derived from ordinal spatial relations and ordinal temporal relations without needing any measures at all. At one point in time automobile A was behind automobile B, and at a later point in time automobile A was ahead of automobile B. This is

---

* In order to show how careful one must be, before qualifying something as "a primitive intuition," let us for a moment consider the meaning of the concept of "overtaking" or "passing." Even here, although it is clear that we have no coordination of measured space and measured time, we might say that we have a coordination of a temporal order and a spatial order. Indeed, what does "overtaking" or "passing" really mean?

1. In a first moment m_1, object A follows object B.
2. In a second moment m_2, A and B are on the same level.
3. In a third moment m_3, A precedes B.

Obviously the temporal series (m_1, m_2, m_3) is coordinated with the spatial series (AB, BA).

sufficient for a child's earliest intuition of speed. It is very easy to show that this intuition of speed precedes any notion of speed in the classical sense as a relationship between a spatial interval and a temporal interval. I should like to mention two experiments that we have done to reveal this precedence of intuition.

In the first experiment we have two tunnels, side by side. One of them is longer than the other, and children have no difficulty seeing this and pointing to the longer one. Then, for each tunnel we have a miniature doll. The dolls are set up to move on tracks at fixed speeds. In the first phase of the experiment we have the dolls enter the tunnels at exactly the same time and emerge from the tunnels at exactly the same time. It is clear that the doll in the longer tunnel must have gone faster, but the unanimous reply from my youngest subjects is that the two dolls moved at the same speeds. The children admit that the dolls went into their tunnels at the same time and came out of them at the same time and that one of them had a much longer tunnel to go through, but nonetheless they assert that the two went at the same speed because they came out at the same time. This is purely an ordinal argument. In the next phase of the experiment we take off the tunnels so that the children see the dolls moving. Once again, the dolls cover the distance in the same time but one of them has a longer distance to cover. This time the very same children say that the doll covering the greater distance goes faster because they can see it pass the other one. They are not coordinating the constant speeds with the different lengths; they are simply reacting

Maybe they understand the question to be 'which exits sooner' (speed = tunnels traversed per unit of time).

to the fact that one of the dolls overtakes the other. In the third phase of the experiment we put the tunnels back over the tracks and repeat the first phase. A great number of our 4- and 5-year-old subjects go right back to what they said in the first phase, namely, that the two dolls go at the same speed because they come out at the same time. Even if we remind them of the second phase, in which they said that one was going faster than the other, they will reply that yes, they remember that, but now the dolls are going at the same speed because they are coming out at the same time.

In another experiment, which is very simple to do, we have two concentric tracks on which, let us say, cyclists are moving. Children will recognize that the outside track covers a longer distance than the inside track. We have the cyclists going around the tracks side by side, so that they get back to the point of departure at the same time. Once again, the children will say that the cyclists travel at the same speed because they return to the same spot at the same time. The fact that the outside track is a longer distance, that the outside cyclist has farther to go, is simply irrelevant to these children's judgments of speed. The only thing that is pertinent to their definition of speed is overtaking, and since the cyclists remain side by side no overtaking occurs. Their judgments of speed are clearly not based on any relationship between length of space on the one hand, and the length of time taken to cover that space on the other hand. This points the way out of the vicious circle, it seems to me, since we see here a notion of speed that is quite different from the classical relation of a measure of space related to a measure of time.

Before dealing with other aspects of the notion of speed and with some aspects of the notion of time, I should like to consider here some experiments in the perception of speed. It is quite clear that we can perceive that one object is going fast or slow, even though it is not passing another object or being passed by another object. We do not have to compare one moving automobile with another in order to see that the automobile is going fast or slow. Upon what is a judgment of this kind based? In an attempt to answer this question we have studied the perception of speed. We have worked with both adults and children, since perception changes much less with age than intelligence does. I shall start by citing the work of an American psychologist, Brown, who studied this subject some time ago and attempted to show that our perceptions of speed result from a relationship between our perceptions of space and our perceptions of time, that is, our subjective impressions of space and our subjective impressions of time. This, of course, is the opposite of what I have been maintaining, and I should like to tell you about a few of our experimental findings that are relevant to this debate.

There is a classical perceptual illusion upon which we based a number of our experiments. The subject looks at a line along which an object moves at a constant speed from left to right. The left half of the line is broken up by small vertical marks, whereas the right half is clear. It is a general perceptual phenomenon that the moving object appears to go faster when it is passing the vertical crosslines than it does during the motion across the other half of the line. With this same experimental situation we can ask subjects to judge not only the speed but also the length of time required for the mobile to move across the left-hand half, compared with the length of time to move across the right half, and we can ask for their judgments on the lengths of the two parts of the line. We do not tell them that half of the line is crossed with vertical marks and the other half is not; we simply ask their judgments on the relative lengths of the crossed part and the open part. In this way we can determine whether or not Brown's position is justified. We can see whether a given subject's judgments of speed, spatial distance, and time interval are consistent with the relationship that speed equals space over time. We carried out these experiments with adult subjects, and, of course, we saw the subjects in three different sessions. No one subject judged time and speed, or distance and speed, in the same session. Nevertheless, when we compared the judgments made by each subject, we found that 60 per cent of the subjects were inconsistent in their judgments. A subject might say, for instance, that it took the same amount of time for the moving object to cross the left-hand part as to cross the right-hand part. In another session he might say that the left-hand part was a shorter distance than the right-hand part. And in still another session he might say that the mobile went faster over the left-hand part than over the right-hand part. These are clearly incompatible judgments if he is operating on the relationship of speed equal to space divided by time. In children the number of inconsistencies was even more marked: closer to 75 or 80 per cent of the children were inconsistent. At any rate it is quite clear that the results for both adults and children do not accord with Brown's point of view.

We are obliged, then, to look for another hypothesis to explain our perception of speed. My hypothesis is that the perception of speed is based on the same sort of ordinal relationships as the notion of speed. I think that we can find this to be the case in three different kinds of situations.

In the first situation there are two moving objects and one of them passes the other. In point of fact, in our experimentation we have found that there is an illusion of acceleration in the speed of one moving object at the moment that it passes the other. Overtaking thus seems to play a role not only in our intuition of speed but also in our very perception of it.

In the second situation there is but a single moving object, and it would seem difficult here to find where the ordinal relationship of overtaking comes in. But in this situation let us say that our eyes are free to move as they will. Hence, in fact, there are once again two moving things—on the one hand the object that we are looking at, and on the other hand our eyes. If we consider, for example, the experiment just discussed, in which an object moves across a line and encounters vertical crosslines as it goes, we see that while the eye follows that moving object it stops for a tiny moment at each of the crosslines, and during this very brief stop the object moves on ahead so that the movement of the eye always seems to be having to catch up from behind. This would explain why the object seems to be moving faster over the crosslined part of the line than over the other part.

The third situation is one in which there is again only one moving object and in which our eyes remain fixed, looking at a fixed point. I can fix my gaze on that "No Smoking" sign, for instance, and without moving my eyes I can tell, more or less, whether a person who walks in front of it is going fast or slow. Once again, we seem hard put to find a case of one moving object overtaking another to serve as a basis for our judgment of speed. In this case, as the mobile moves across the field of vision, it excites simultaneously a certain number of retinal cells. I call the collection of cells that are stimulated at any given moment the train of excitation, and this is the source of the two moving objects in this situation, namely, the first cell in the train of excitation and the last cell in the train—the locomotive and the caboose, if you will. The faster an object moves across our field of vision, the greater is the distance between the first cell and the last cell, and it is this increase in distance that leads to our judgment of an increase in speed. This same explanation, by the way, accounts for the fact that in this third situation, when a mobile moves across our field of vision and our eyes are fixed, it seems to speed up as it passes the foveal region. The cells are denser in this region, so that as the mobile reaches there, there are more cells between the beginning and the end of the excitation train. This is what gives rise to the impression that the object seems to speed up at this point.

I have only two final remarks to make on the ordinal nature of our perceptions and intuitions of speed. One concerns the physiological work of Letvin at the Massachusetts Institute of Technology. Letvin has been studying, among other things, the retinal sensitivity of frogs and has found that there is a primitive perception of speed. He finds no such primitive perception of time, however.

My second remark refers to the work of two French physicists. They were attempting to establish a new axiomatization of physics that could serve as a basis of relativity physics. Among other things they wanted to be able to avoid the problem of the vicious circle in the notions of speed and time. These two physicists had the great merit to look into psychological studies of the notion and the perception of speed and time, and they came upon our work. They found in our hypothesis of the ordinal notion of speed a way to introduce into their formal structure a notion of speed that was independent of temporal duration, and this is what enabled them to escape from the vicious circle. It is interesting to me that in this way the influences from one branch of investigation to another have come full circle. It was the author of the theory of relativity who suggested to us our work, which in turn proved to be useful to other physicists in building the axiomatic basis of the theory of relativity.

I should like now to consider the notion of time. Whereas

we have seen that there is a primitive intuition of speed, this is not at all the case with time; the notion of time is an intellectual construction. It is a relationship between an action—something which gets done—and the speed with which it is done.

It is very easy to show that in the development of the notion of time in small children this relationship is not a primitive intuition. Judgments of time are based on how much has been accomplished or on how fast an action has taken place, without the two necessarily having been put into a relationship with one another. Let us look at the development of the notion of simultaneity, for instance. In one of our experiments the experimenter has two little dolls, one in each hand, that walk along the table side by side (they do not actually walk; they go in hops, tapping the table together at the end of each hop). The child says go; the two dolls start off at exactly the same time and the same speed. The child says stop, and the two dolls stop, once again side by side having gone exactly the same distance. In this situation children have no problem in admitting that the dolls started at the same time and stopped at the same time. But if we change the situation slightly, so that one of the dolls has a slightly longer hop each time than the other, then, when the child says stop, one doll will be farther along than the other. In this situation the child will agree that the dolls started at the same time, but he will deny that they stopped at the same time. He will say that one stopped first; it did not go as far. We can then ask him, "When it stopped, was the other one still going?" And he will say no. Then we will ask him, "When the other one stopped, was this one still going?" And he will say no again. This is not, then, a question of a perceptual illusion. Finally, we will ask again, "Then did they stop at the same time?" The child will still say, "No, they did not stop at the same time because this one did not get as far." The notion of simultaneity—two things happening at the same time—simply does not make sense for these children when it refers to two qualitatively different motions. It makes sense for two qualitatively similar motions taking place at the same speed, as in the first situation described, but when two different kinds of motions are involved it simply makes no sense. There is no primitive intuition of simultaneity, and two movements are qualitatively different. This is going to require an intellectual construction.

Slightly older children will admit that the two dolls stopped at the same time, but they will still have trouble with the question of whether the dolls walked or moved for the same length of time; that is, they will have trouble with questions of the time interval or time duration. They will say that the dolls started at the same time, and they stopped at the same time, but that one walked a longer time because it got farther. It is very clear here that the notion of time is being based on the amount of action carried out or the speed at which the action is carried out, but these two have not been put into a relationship with one another to give rise to a consistent notion of temporal duration. A period of time cannot be dissociated from what is accomplished during this period.

Another experiment is even simpler for studying these same notions. Using a Y-shaped tube, we can attach the stem of the tube to a tap of water so that there is an equal flow out of each of the branches. Each branch can flow into a separate container. If the two containers are the same size and shape and we question the children about the water flow when we turn on the tap, the children will admit that the water starts to flow into the two containers at the same time, stops flowing at the same time, and has flowed the same length of time into both of them. However, if we have differently shaped containers, so that the water rises higher in one than in the other after a given amount of time, the children have the same problems again, saying that the water ran for a longer time into the container where it has risen higher.

In many of these cases we can point out to the child the time consistencies involved by giving him a watch or other time instrument, but when we do this we find that it does not help at all, because these children have no notion of the constancy of the speed of the measuring instrument. As they see the situation, if the watch goes farther one time than another, the reason may very well be because it is going faster at that time. Or if sand runs through two egg timers in the same length of time but the child really thinks the two events took different lengths of time, he simply maintains that the sand went faster in one egg timer than in the other, or it went faster in the same egg timer one time than it went in another. There is simply no notion that the speed remains constant in these instruments.

As a final remark on the preoperational notions of time I should like to mention that some children think that faster means a longer period of time. For instance, we ask them how long it takes them to walk to school. A child may say a quarter of an hour. Then we will ask him whether, if he runs to school, it takes longer than a quarter of an hour or shorter than a quarter of an hour. The child will often say that it takes longer than a quarter of an hour because, once again, he has been unable to make the proper relationship between the amount of work done and the speed at which it is done in order to produce from this the relative amount of time that passes. It is as if he is reasoning in this way: faster means more gets done, and getting more done means spending more time.

A word or two about subjective time or psychological time may also be helpful. Offhand it may seem that this is quite a different question, since we appear to have a direct impression of subjective time, but on looking more closely we see that, in fact, the same relationship is in play here. Our subjective impressions of time depend, on the one hand, on the actions that we are carrying out or the amount of work that is accomplished and, on the other hand, on the speed at which the work is being accomplished. Why, for example, does time seem shorter when we are doing something that interests us? The answer is a very simple one. Dewey, a long time ago, and Claparede, too, have pointed out that interest reinforces or accelerates the speed with which work is done.

In this area I find myself in partial disagreement (but

only partially) with my colleague Fraisse, a specialist in the psychology of time. He believes that the subjective impression of time is a function of the number of events or the number of changes that the subject notices. In other words, the more varied are the contents of our experience, the longer the time seems to be. What seems to be missing behind this hypothesis is the notion of the number of events in relationship to a fixed unit of time, that is, a notion of the frequency of the events. I think that this element of frequency, which is a form of speed, is hidden within Fraisse's framework. Let us look at the following experiment, which Fraisse did first and we repeated. Various pictures are shown to children during 1 minute of time. In one situation they are shown sixteen pictures in a minute; in another situation, thirty-two pictures in a minute. Small children under the age of 7 years judge that the time is longer when they see thirty-two pictures in a minute than when they see sixteen pictures. This seems to support Fraisse's hypothesis, but if we do the same experiment with slightly older children—7- or 8-year-olds—we find the reverse judgment. These children seem to judge that the time is shorter in the situation when they see thirty-two pictures. It seems quite clear that here the speed of the events must be playing a role in the judgment of them, and, in fact, it seems to be playing the deciding role.

I shall conclude my remarks on the notion of time by saying that it requires a construction—an intellectual construction on the part of children—based on operations that are parallel to those involved in logical, mathematical thinking. Three kinds of operations are involved in the notion of time. First of all, there are the operations of seriation, of ordering events in time: B comes after A, C comes after B, D comes after C, etc. Second, there are the operations that are similar to the operations of class inclusion: if event B follows event A and event C follows event B, then we must be able to conclude operationally that time interval AC is longer than time interval AB. This corresponds in the logic of classes to the notion that the whole is greater than a part, or a whole class—the total class—is greater than a subclass. And, finally, we have the operations of measurement of time, which are the synthesis of the two other kinds of operations, just as the operations involving number are the synthesis of operations of ordering and of classification.*

* It may be asked how it is possible for somebody who defines intelli gence by means of reversible structures to make intelligible the notion of time, precisely characterized by pure irreversibility. Our answer is simple: this physically irreversible time is, in thought, reversible (we can go back and forth, from present to past and from past to present) by means of our reversible interiorized operations.

Conclusion

T HESE few examples may clarify why I consider the main problem of genetic epistemology to be the explanation of the construction of novelties in the development of knowledge. From the empiricist point of view, a "discovery" is new for the person who makes it, but what is discovered was already in existence in external reality and there is therefore no construction of new realities. The nativist or apriorist maintains that the forms of knowledge are predetermined inside the subject and thus again, strictly speaking, there can be no novelty. By contrast, for the genetic epistemologist, knowledge results from continuous construction, since in each act of understanding, some degree of invention is involved; in development, the passage from one stage to the next is always characterized by the formation of new structures which did not exist before, either in the external world or in the subject's mind. The central problem of genetic

Tacit knowledge

epistemology concerns the mechanism of this construction of novelties which creates the need for the explanatory factors which we call *reflexive abstraction* and *self-regulation*. However, these factors have furnished only global explanations. A great deal of work remains to be done in order to clarify this fundamental process of intellectual creation, which is found at all the levels of cognition, from those of earliest childhood to those culminating in the most remarkable of scientific inventions.

INDEX

Abstraction, logical and mathematical, 16
Abstraction, simple and reflective, 17
Actions: types of, 18; logic of, 42
Additive coordination, 18
Additive group, 22, 25
Algebraic structure, 3, 24, 25; in children's thinking, 27, 38
Apparent motion, 55
Axiomatic systems, 11, 69

Bergson, Henri, 6
Beth, Evert W.: quoted, 12
Binet, Alfred, 31
Blind children, 46
Bloomfield, Leonard, 9
Bohr, Niels, 3
Bourbaki mathematicians, 3, 23, 51
Boyat, Gilbert, 54, 56
Bruner, Jerome, 53

Cantor, Georg F. L. P., 4, 34
Categories, notion of, 3
Children: development of notion of number in, 5; origins of notion
    of time and of simultaneity in, 7; logical behavior, 9; develop-
    ment of knowledge studied in, 13 ff.; thinking structures of, 26;
    reasoning on basis of transitivity, 30; development of thinking,
    31; deaf and dumb, 45; blind, 46; Sinclair experiment, 48 ff.;
    type of thinking and logical reasoning shown by, 50 ff.;
    development of notion of speed, 62; development of notion of
    time, 70

79

Chomsky, Avram N., 9, 47
Claparede, Edouard, 73
Classification, logic of: in children's thinking, 27
Classification, operational structure of, 28
Class inclusion, structure of, 28, 75
Commutativity, 17
Conservation, 43, 48, 51, 52, 53, 54
Coordinated actions, 18, 21, 42
Curvilinear figures: as distinct from rectilinear figures, 33

Deaf and dumb children, 45
de Broglie, Louis, 3
Determinism, 3
Detour behavior, 44
Dewey, John, 73
Dieudonne, J., 26
Duration (term), 61

Eilenberg, Samuel, 3
Einstein, Albert, 6, 69
Empiricism, 16
Epistemology: traditional philosophical view of, 1 ff.; relevance of
    psychological factors in, 7 ff.
Epistemology, genetic: objections to, 2 ff.; importance of
    psychological research in, 9; concerned with formation and
    meaning of knowledge, 12; fundamental hypothesis of, 13;
    central problem of, 77
Equivalent classes, 5, 36
Euclidian metric geometry, 30

Figurative aspect of thinking, 14
Finite number series, 4
Formalization, logical, 10 ff.
Fraisse, Roland, 74
Functions, 51, 52
Furth, Hans, 46

Geometry, development of, 30
Geometry, projective, 30
Geometry, pure: as science of independent space, 60
Gödel's theorem, 11
Group of displacements, 25, 44

Ideas, development of, 2
Identity, notion of, 52, 53, 54, 56, 57
Imitation: as figurative function of thinking, 14; deferred, 45
Indeterminism, 3
Intellectual construction, 7, 70, 74
Intellectual development, mathematical structures and
        language in, 41
Intelligence, in deaf and dumb children, 45
Intelligence, practical, 41, 44
Interiorized imitation, 14
Intuition, articulated, 50
Intuition, primitive, 7, 62
Inversion: reversibility by, 22
Irreversibility of time, 60, 75

Knowledge: epistemology as study of, 2; factors in understanding
        nature of, 4, 7; states of, 12 ff.; active quality of, 15; concrete
        quality of physical, 16; logical mathematical, 17; of space and
        time, 61; development of, 77

Language: as based on reason, 10; relationship between
        mathematical structures and, 41 ff.; practical intelligence
        evident before use of, 44 ff.
Language development: in relation to logical structures, 9
Laws of transformation, 23
Leurendau, Monique, 33
Linguistic level of children: in relation to operational level, 48
Linguistic structures, 8
Logic: linguistic view of, 9; many kinds of logics, 10 ff.
Logical positivism, 8, 48
Logical reality, 8

Logical structures: development of, in children, 14; abstract
     quality, 15; basis of, 18; development of, in deaf and dumb
     children, 46

Maritain, Jacques, 6
Mathematical and Psychological Epistemology (Piaget and Beth),
     12
Mathematical groups, 25
Mathematical reality, 8
Mathematical structures, 3, 22; abstract quality, 15; relationship
     between language and, 41 ff.; relationship between sensory-
     motor activities and, 41 ff.
Mathematics: fundamental structures of, 3
McCulloch, W. S., 19
Mental imagery, 14, 45
Mental operations, 21 ff.
Michaelson-Morley experiment, 6
Microphysics: changing state of knowledge in, 2
Mother structures of Bourbaki mathematicians, 23 ff., 26, 33
Motion, relative, 39
Movement, notion of: hypothesis of, 59 ff.

Natural numbers, notion of, 5
Negation: reversibility by, 22, 28, 39
Number: as a synthesis of class inclusion and relationships of
     order, 38
Number, notion of, development of, in children's thinking, 5, 34,
     38; on basis of one-to-one correspondence, 37
Number theories: bases for, 39

Oleron, Pierre, 46
One-to-one correspondence, 4, 34 ff., 43
Operational level of children: in relation to linguistic level, 48
Operations: development of, 2, 43; defined, 21 ff.
Operative aspect of thinking, 14
Optical correspondence, 34
Order, logic of, 42

Order structure, 25; in children's thinking, 28, 38
Order (term), 61
Ordinal coordination, 18
Overtaking, concept of, 62

Passing, phenomenon of, 62
Perception: as figurative function of thinking, 14
Permanence, notion of, 43
Physics: changing state of knowledge in, 2
Pinard, Adrien, 33
Poincaré, Henri, 7
Position, changes of, 59
Principia Mathematica (Russell and Whitehead), 35
Propositions, logic of, 25, 39
Psychological factors : in epistemological analysis, 7 ff.

Rationalism, 9
Reality, logical and mathematical, 8
Reciprocity: reversibility by, 22, 25, 29, 30, 39
Rectilinear figures: as distinct from curvilinear figures, 33
Reflexive abstraction, 78
Relativity theory, 7, 60
Representation, 45
Reversibility: in sensory-motor intelligence, 43, 44; of movement
    in space, 60
Reversible actions, 21, 29
Russell, Bertrand, 5, 35 ff.

Scalers, 49
Schemes, logic of, 42
Self-regulation, 78
Semilogic in children, 50, 52
Semiotic function, 45
Sensory-motor activities: relationship between mathematical
    structures and, 41 ff.
Sensory-motor intelligence: internalizing of, 45; notion of identity
    in, 52

Sequential coordination, 18
Seriation, structure of, 28
Set theory, 4
Simultaneity, notion of, 6 ff., 70, 71
Sinclair, Hermine, 48
Space: as coordination of changes of position, 59; parallel between
    time and, 60
Speed: notion of, 7, 59 ff.; intuition of, 62 ff.; perception of, 65 ff.
Stroboscopic motion, 55
Structuralist school of mathematics, 3
Structure: defined, 22 ff.
Structure of ordering, 3
Structures: hypothesis of innate, 47
Systems of transformations, 14 ff.

Temporal notions, 61
Thinking without Language (Furth), 46
Thought: as a process, 2 ff.; figurative and operative aspects, 14 ff.;
    hypothesis of, 18
Time: notion of, 7, 59 ff.; parallel between space and, 60;
    irreversibility of, 60, 75; an intellectual construction, 70, 74;
    subjective, psychological time, 73; kinds of operations involved
    in, 75
Topological geometry, 31
Topological intuitions, 31
Topological structure, 3, 25; in children's thinking, 30
Train of excitation, 68
Transfinite numbers, 34
Transitivity: reasoning on basis of, 30

Unity, arithmetic, 37
Universal time, Newtonian notion of, 6

Vectors, 49

Whitehead, Alfred North, 5, 35 ff.

Norton Paperbacks
PSYCHIATRY AND PSYCHOLOGY

Adorno, T. W. et al.   *The Authoritarian Personality.*

Alexander, Franz.   *Fundamentals of Psychoanalysis.*

Alexander, Franz.   *Psychosomatic Medicine.*

Bruner, Jerome S.   *The Relevance of Education.*

Bruner, Jerome S.   *Toward a Theory of Instruction.*

Cannon, Walter B.   *The Wisdom of the Body.*

Erikson, Erik H.   *Childhood and Society.*

Erikson, Erik H.   *Gandhi's Truth.*

Erikson, Erik H.   *Identity: Youth and Crisis.*

Erikson, Erik H.   *Insight and Responsibility.*

Erikson, Erik H.   *Young Man Luther.*

Ferenczi, Sandor.   *Thalassa: A Theory of Genitality.*

Field, M. J.   *Search for Security: An Ethno-Psychiatric Study of Rural Ghana.*

Freud, Sigmund.   *An Autobiographical Study.*

Freud, Sigmund.   *Beyond the Pleasure Principle.*

Freud, Sigmund.   *Civilization and its Discontents.*

Freud, Sigmund.   *The Ego and the Id.*

Freud, Sigmund.   *Group Psychology and the Analysis of the Ego.*

Freud, Sigmund.   *Jokes and Their Relation to the Unconscious.*

Freud, Sigmund.   *Leonardo da Vinci and a Memory of His Childhood.*

Freud, Sigmund.   *New Introductory Lectures on Psychoanalysis.*

Freud, Sigmund.   *On Dreams.*

Freud, Sigmund.   *On the History of the Psycho-Analytic Movement.*

Freud, Sigmund.   *An Outline of Psycho-Analysis Rev. Ed.*

Freud, Sigmund.   *The Problem of Anxiety.*

Freud, Sigmund.   *The Psychopathology of Everyday Life.*

Freud, Sigmund.   *The Question of Lay Analysis.*

Freud, Sigmund.   *Totem and Taboo.*

Horney, Karen (Ed.)   *Are You Considering Psychoanalysis?*

Horney, Karen.   *Feminine Psychology.*

Horney, Karen.   *Neurosis and Human Growth.*

Horney, Karen.   *The Neurotic Personality of Our Time.*

Horney, Karen.   *New Ways in Psychoanalysis.*

Horney, Karen.   *Our Inner Conflicts.*

Horney, Karen.   *Self-Analysis.*

Inhelder, Bärbel and Jean Piaget.   *The Early Growth of Logic in the Child.*

James, William.   *Talks to Teachers.*

Kasanin, J. S.   *Language and Thought in Schizophrenia.*

Kelly, George A.   *A Theory of Personality.*

Klein, Melanie and Joan Riviere.   *Love, Hate and Reparation.*

Levy, David M.   *Maternal Overprotection.*

Lifton, Robert Jay.   *Thought Reform and the Psychology of Totalism.*

Piaget, Jean. *The Child's Conception of Number.*
Piaget, Jean. *Genetic Epistemology.*
Piaget, Jean. *The Origins of Intelligence in Children.*
Piaget, Jean. *Play, Dreams and Imitation in Childhood.*
Piaget, Jean and Bärbel Inhelder. *The Child's Conception of Space.*
Piers, Gerhart and Milton B. Singer. *Shame and Guilt.*
Ruesch, Jurgen. *Disturbed Communication.*
Ruesch, Jurgen. *Therapeutic Communication.*
Ruesch, Jurgen and Gregory Bateson. *Communication: The Social Matrix of Psychiatry.*
Schein, Edgar et al. *Coercive Persuasion.*
Sullivan, Harry Stack. *Clinical Studies in Psychiatry.*
Sullivan, Harry Stack. *Conceptions of Modern Psychiatry.*
Sullivan, Harry Stack. *The Fusion of Psychiatry and Social Science.*
Sullivan, Harry Stack. *The Interpersonal Theory of Psychiatry.*
Sullivan, Harry Stack. *The Psychiatric Interview.*
Sullivan, Harry Stack. *Schizophrenia as a Human Process.*
Walter, W. Grey. *The Living Brain.*
Watson, John B. *Behaviorism.*
Wheelis, Allen. *The Quest for Identity.*
Zilboorg, Gregory. *A History of Medical Psychology.*

Liveright Paperbacks

Balint, Michael. *Problems of Human Pleasure and Behavior.*
Bauer, Bernhard A. *Woman and Love.*
Bergler, Edmund. *Curable and Incurable Neurotics: Problems of "Neurotic" versus "Malignant" Psychic Masochism.*
Bergler, Edmund. *Parents Not Guilty of Their Children's Neuroses.*
Coles, Robert et al. *Drugs and Youth: Medical, Psychiatric, and Legal Facts.*
Dewey, John. *The Sources of a Science of Education.*
Dunlap, Knight. *Habits: Their Making and Unmaking.*
Featherstone, Joseph. *Schools Where Children Learn.*
Gutheil, Emil A. *The Handbook of Dream Analysis.*
Gutheil, Emil A. *Music and Your Emotions.*
Jones, Ernest. *On the Nightmare.*
Köhler, Wolfgang. *Dynamics in Psychology.*
Köhler, Wolfgang. *Gestalt Psychology.*
Köhler, Wolfgang. *The Selected Papers of Wolfgang Köhler.*
Russell, Bertrand. *Education and the Good Life.*
Stekel, Wilhelm. *Impotence in the Male.*
Stekel, Wilhelm. *Sexual Aberrations: The Phenomena of Fetishism in Relation to Sex.*